Australian
GEOGRAPHIC
DISCOVER

ROCKS & FOSSILS
CONTENTS

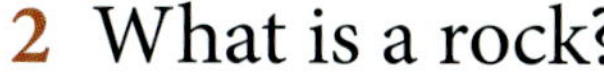

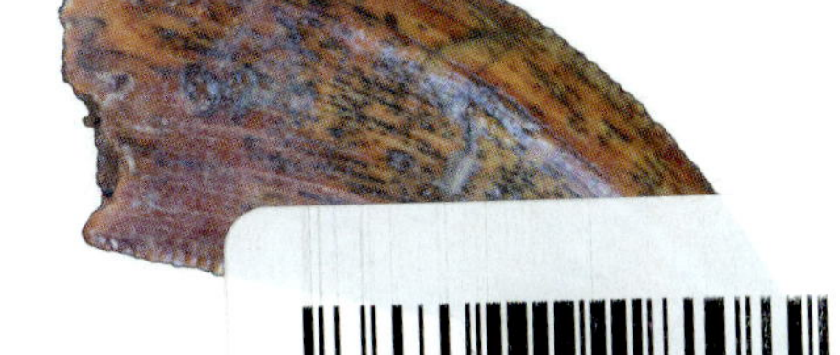

DOME ROCK, BOOLCOOMATTA, SA

LOST CITY, LIMMEN NATIONAL PARK, NT

Rocks are solid objects made up of **MINERALS**. They are found in or on the ground – the crust that covers the Earth. A long time ago, the Earth was covered in boiling, liquid **MAGMA,** and when this cooled, it formed the first rocks. Some of the flat layers of rock that formed were pushed up and crinkled, forming mountains and valleys.

REMARKABLE ROCKS, KANGAROO ISLAND, SA

Rocks often settle on the ground in layers. These layers, made of different minerals, can be different colours, making the rock look striped. There are three types of rock – igneous, metamorphic and sedimentary.

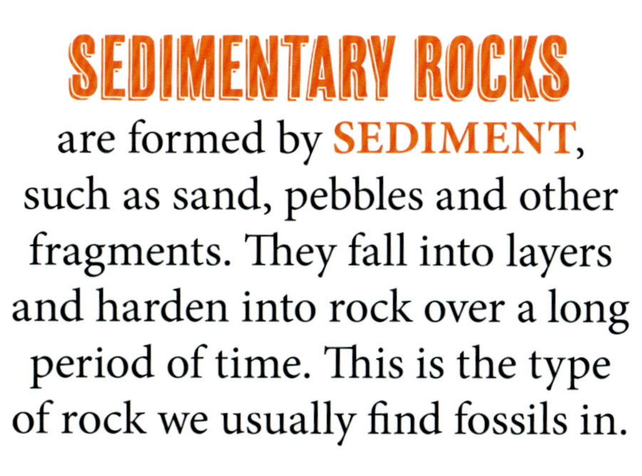

## SEDIMENTARY ROCKS

are formed by **SEDIMENT**, such as sand, pebbles and other fragments. They fall into layers and harden into rock over a long period of time. This is the type of rock we usually find fossils in.

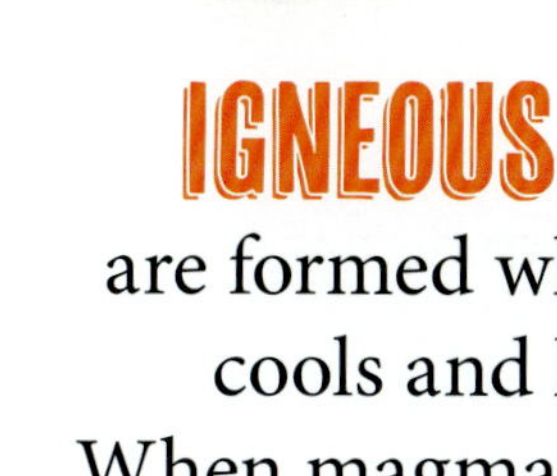


## METAMORPHIC ROCKS

are formed out of minerals deep below the Earth's surface. The intense heat and pressure below the crust transform minerals into hard rocks like marble and slate.

## IGNEOUS ROCKS

are formed when magma cools and hardens. When magma erupts from volcanoes, it's called lava. The faster it cools, the shinier and more like glass it will look.

# SCIENTISTS WHO ROCK

People who study rocks are called geologists. Geologists study minerals and the processes and history that have shaped rocks. This can help explain how landforms like mountains, valleys and rivers got their shapes.

## DID YOU KNOW?

The word 'geology' means the study of (logy) the Earth (geo). Do you know any other words that begin with 'geo'?

LOCH ARD GORGE, VIC

DONNA CAVE, QLD

GOLD

In the 1700s, scientists began discovering that rock layers can explain what happened in an area thousands of years ago. Many parts of Australia, for example, are covered in limestone, which tells geologists that at some point in time these areas were under the sea. Limestone is made from the broken-down bodies of dead sea creatures that have settled on the sea floor.

Geologists can also help determine where there are likely to be certain minerals in the ground, such as gold or gemstones.

## GOING UNDERGROUND

There are often caves around areas with lots of limestone, as parts of the mineral may wear away, creating holes. Australia has one of the biggest stretches of limestone in the world, a region across Western Australia and South Australia called the Nullarbor. The Nullarbor is full of caves!

## DOLERITE

At the southern end of Tasmania stand huge columns of dolerite. Formed by cooling magma, these hexagonal columns rise hundreds of metres out of the sea.

# VOLCANIC ROCKS

Once **LAVA** erupts from a volcano, it cools and solidifies into igneous rock. Igneous rock can be very different depending on the area it is in, how deep the volcano is, how hot the lava in it is and how fast or slowly the rock cools. Vulcanologists (people who study volcanoes) and geologists examine these rocks to tell whether an area has been formed through volcanic activity and if there are currently, or once were, volcanoes in the area. This can tell geologists all sorts of things, like how the Earth's crust has moved over time.

## LAVA TUBES

In far north Queensland, there are cave systems formed by lava! Beginning 190,000 years ago, Undara volcano erupted with a long, seething gush of lava. It filled nearby valleys to the brim, and although the surface cooled into a hard crust, liquid lava kept surging below. More than 160km of lava tubes, tunnels and arches were created.

## DID YOU KNOW?

Australia has no active volcanoes on the mainland, but there are some active ones on Australian-owned Antarctic islands.

# IDENTIFYING MINERALS

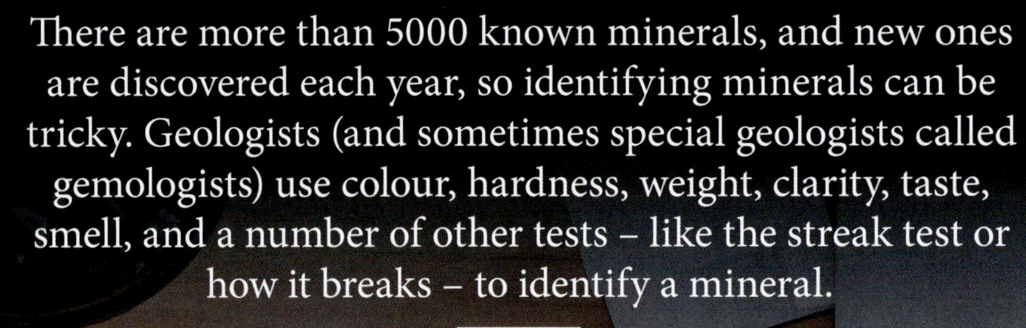

There are more than 5000 known minerals, and new ones are discovered each year, so identifying minerals can be tricky. Geologists (and sometimes special geologists called gemologists) use colour, hardness, weight, clarity, taste, smell, and a number of other tests – like the streak test or how it breaks – to identify a mineral.

## THE STREAK TEST

A streak test is where geologists rub the mineral on a porcelain plate. The colour of the streak left by the mineral is sometimes different from the colour of the mineral itself.

## DID YOU KNOW?

To test the hardness of different minerals, you can scratch them against one another. The rock that scratches the other is the harder of the two!

## LOOK CLOSELY

Often, gems dug up from underground are hard to identify. They might be coated with a layer of dirt or clay, covering up obvious signs. Rocks brought up from an opal mine, for instance, need to be washed and carefully sorted.

RUBY

ROSE QUARTZ

BLUE AGATE

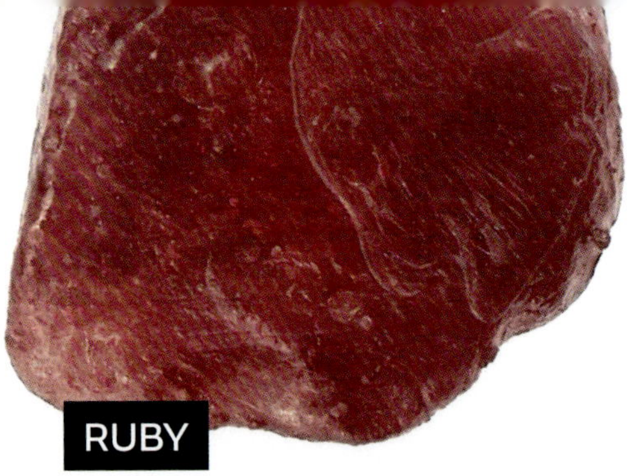

LAPIS LAZULI

OPAL

# PRECIOUS GEMS

A gem is a hard mineral that is desirable for its beauty and valuable for its rarity. People often make gems into jewellery and have traded them around the world like currency for centuries. Miners usually dig them up, although some gems can be found on top of the ground.

EMERALD

GARNET

SAPPHIRE

## PRECIOUS STONES

Diamonds, rubies, sapphires and emeralds are worth money because they are clearer and harder than most stones. Most other gemstones are called semi-precious stones.

## DID YOU KNOW?

Amethyst has been used for decoration and jewellery since the time of the ancient Egyptians!

AMETHYST

## PEARLS

Natural pearls form inside oysters! If a foreign object, such as a grain of sand, slips into the shell, the oyster will cover it with nacre, a mineral. The oyster adds layer upon layer until the gem is formed.

# FOSSICKING IN AUSTRALIA

When European settlers went looking for gems and minerals in Australia, they developed systems to search through sand, dirt and mud. They would often use pans and water to sift through the sediment, spotting small specks of gems and minerals as they went.

## NOODLING

'Noodling' is a term used to describe fossicking for opal. It is a favourite pastime in Coober Pedy, South Australia, where an entire community is based underground near the opal mines.

Many Australians fossick for fun, using it as a reason to explore out-of-the-way places like central Queensland, where sapphires and emeralds can be found, among other gems.

# ULUṞU AND OTHER BIG ROCKS

Uluṟu and Kata Tjuṯa in the Northern Territory were created at the same time more than 500 million years ago. They are the remains of a mountain range that had little vegetation to protect it from winds and rain that broke it down, carrying the soil and rock away. These mountain ranges eroded away almost completely. The rocks that stick up now as Uluṟu and Kata Tjuṯa are all that remain. They were folded tightly when the mountains originally crinkled up, and that may be why the rock itself is so hard, resistant to erosion and uniform in colour.

## DID YOU KNOW?

Their red colour comes from iron in the rock. It turns red like rust when exposed to air and water. Originally Uluṟu and Kata Tjuṯa would both have been grey.

## WAVE ROCK

Found in the south-eastern corner of Western Australia, Wave Rock is part of a large granite formation known as Hyden Rock. The 'wave' is about 14m tall, and 110m long!

## BOTTOM TO TOP

The rocks of the Bungle Bungle Range in Purnululu National Park, Western Australia, formed in a similar way to Uluṟu. They have banded stripes of colour because they used to be at the bottom of a river bed. Over aeons, layers of sediment settled on the bottom of the river and eventually compressed into sandstone and lifted up to form a mountain range.

# WHAT IS A FOSSIL?

ELASMOSAUR FOSSIL

FOSSILISED SHELLS

Ancient bones and dinosaur skeletons are one type of fossil. Other things can also be fossils, such as footprints.

## DID YOU KNOW?

Ancient animal poo can be turned into a fossil. The scientific name for a poo fossil is a coprolite.

Fossils are the remains of an animal or plant that died long ago. They are rocks that show the shape of a plant, animal, or other organic material. These things are preserved when layers of sediment form into rocks around them. Rock protects fossils from getting too much **OXYGEN**, **BACTERIA** or water, which would make fossils break down.

DROMINICID FOSSIL

AMMONITE FOSSIL

## LARK QUARRY

This conservation park in western central Queensland has lots of dino footprints. For a long time, scientists thought it showed a **STAMPEDE**. Today, it is thought these footprints came from dinosaurs walking in a river.

In June 2009, the fossils of a giant plant-eating dinosaur (*Diamantinasaurus*) were found near Lark Quarry. Longer than three and a half cars, it was nicknamed 'Matilda'. The bones of a smaller dinosaur (*Austrolovenator*) were found tucked below. Nicknamed 'Banjo', it had 30cm-long claws and flesh-slicing teeth.

PALAEONTOLOGISTS LOOKING FOR FOSSILS AT RIVERSLEIGH, QLD

# FOSSIL FIENDS

Palaeontology is the study of the history of life on Earth, based on fossils. Palaeontologists look at fossils and try to figure out what happened when the fossil was alive. Most study specific things – like plants, ancient humans or even very small fossils. For example, a plant palaeontologist might look at fossilised tree rings. You can tell if an area was easy for a tree to live in thousands of years ago by looking at how thick the ring is. If the ring of a tree is thick, it probably means it was an easy place to live and the tree was able to grow rapidly within a year.

As well as dinosaurs, we also find fossils from other **EXTINCT** species of animals that once roamed Australia. This fossil is from *Mayigriphus orbus*, an animal that was a bit like a quoll.

## DID YOU KNOW?

Australian fossils are still being found and described. This is a radius (a forearm bone) of a sauropod found in Winton, Queensland.

SHARK TOOTH FOSSIL

# FOSSILS ARE FASCINATING

SPINOSAURUS TEETH FOSSILS

Scientists love fossils because they tell us all sorts of things about what life was like a long time ago and why we are the way we are today.

VELOCIRAPTOR TOOTH FOSSIL

We can learn all sorts of things from fossils. If an animal had sharp teeth, we know it may have been a meat-eater that needed to snap up and tear apart other animals.

If an animal had flat teeth, it is more likely it ate plants and had to grind them down so it could digest them.

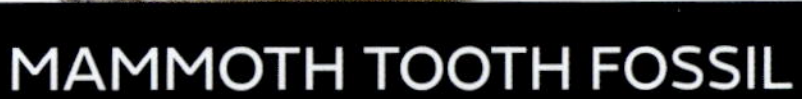

MAMMOTH TOOTH FOSSIL

## AUSTRALIAN FOSSILS

Australia is home to some of the oldest fossils in the world. Tiny fossils found in Western Australia could be up to 3.5 billion years old, and might even be our ancient relatives! Scientists think these organisms were smelly and coloured purple and green. Eventually, these tiny things evolved into bigger creatures, which also became bigger and more complicated until they became the animals and humans we see today.

### DID YOU KNOW?

Some dinosaurs were feathered. Scientists can use these feathers to determine what colour some dinosaurs were!

**ARCHAEOPTERYX** was a bird-like dinosaur that lived around 150 million years ago in what is now southern Germany.

# AUSTRALIA'S FOSSIL SITES

Australia's fossils are found in some of the most hard-to-get parts of the country, from the Otway Ranges at Victoria's southern tip, to the underground opal mines of Lightning Ridge in north-western New South Wales and the plains of central Queensland.

## LIGHTNING RIDGE, NSW

At Lightning Ridge, amazing fossils can be found – fossils of dinosaurs and other ancient creatures made of opal!

## FOSSIL COVE, TAS

Just south of Hobart in Tasmania is a wave-formed rock wall filled with fossils.

## DID YOU KNOW?

*Muttaburrasaurus* is named after the town it was found in – can you spot it on the map?

## BROOME, WA

The dinosaur footprints, a kind of trace fossil, found in Broome are about 130 million years old. The tracks belong to a range of dinosaurs, including big sauropods and smaller theropods.

## GERALDTON, WA

In 1966, some schoolboys found a dinosaur bone near Geraldton that belonged to a speedy meat-eater known as *Ozraptor*.

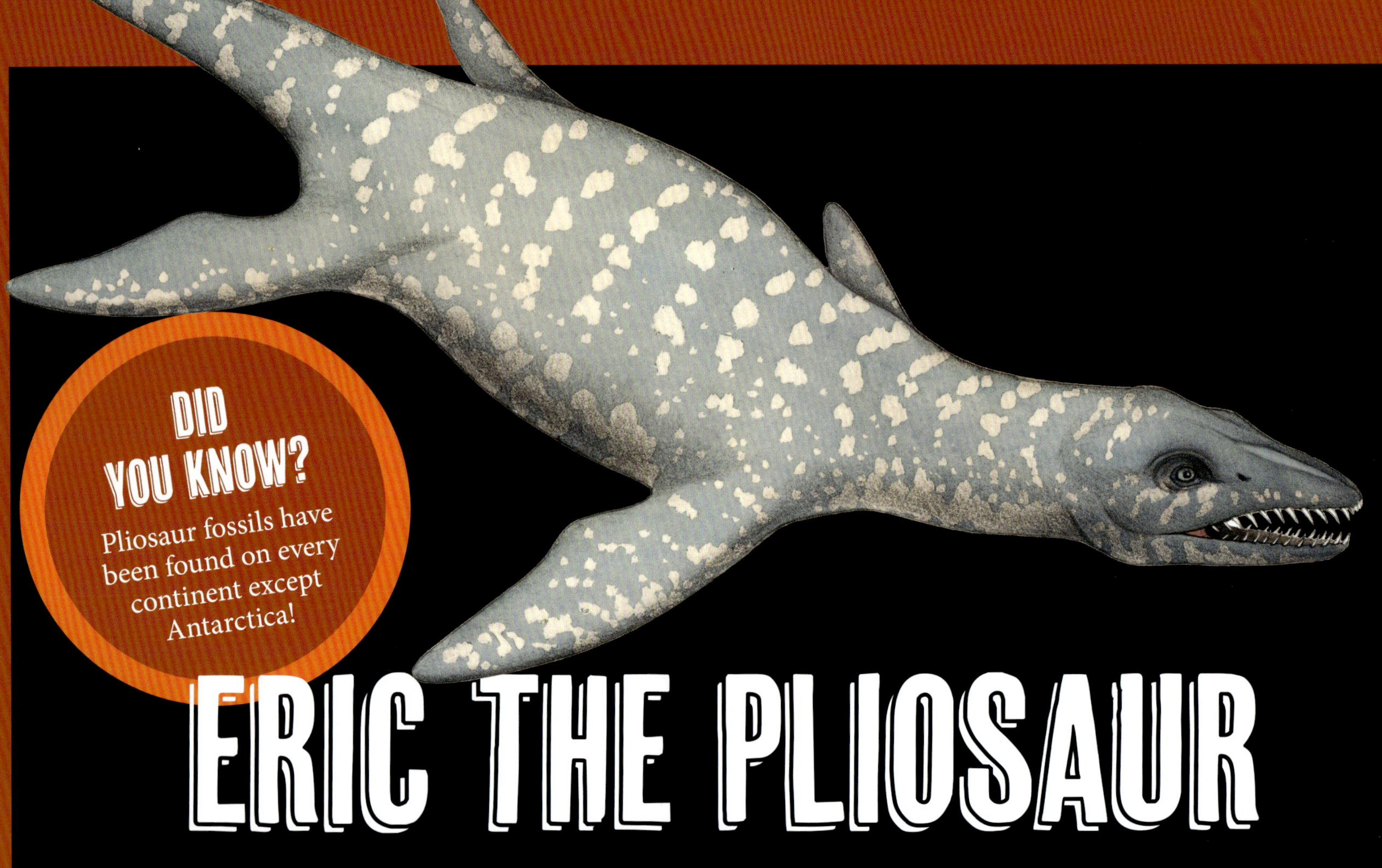

DID YOU KNOW?

Pliosaur fossils have been found on every continent except Antarctica!

# ERIC THE PLIOSAUR

One of the more interesting Australian fossil stories is that of 'Eric' the pliosaur, whose bones were opalised, leaving the fossil with a semi-precious sheen. Pliosaurs were huge reptiles from the Jurassic era that swam in an enormous, shallow inland sea in the middle of Australia. These carnivorous reptiles fed on fish, sharks, dinosaurs and other reptiles and are the ancient relatives of turtles. Pliosaurs could reach 15m long, but Eric was very small, only about the size of a seal.

COOBER PEDY, SA

## DUGOUT

Coober Pedy is famous for the underground homes that more than half of the locals live in.

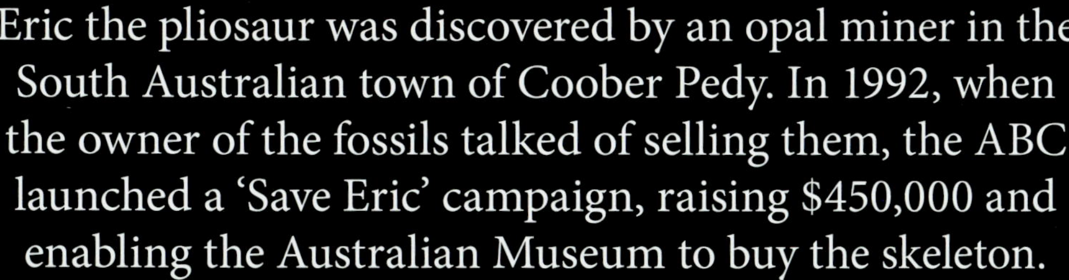

Eric the pliosaur was discovered by an opal miner in the South Australian town of Coober Pedy. In 1992, when the owner of the fossils talked of selling them, the ABC launched a 'Save Eric' campaign, raising $450,000 and enabling the Australian Museum to buy the skeleton.

## SHIMMERING COPIES

Opals are a kind of gemstone. Things can become opalised when something, such as bone or a piece of wood, disintegrates and the right substances seep into the gap, creating a copy over thousands of years. Australia is very rich in this gemstone, producing 95 per cent of the world's precious opal.

# FOSSIL FUELS

Fossil fuels like oil, coal and natural gas come from decayed plant and animal remains that were exposed to heat and pressure in the Earth's crust over hundreds of millions of years. They are often found far underground and are very valuable. Fossil fuels produce electricity and fuel our cars, stoves and heaters.

## HOW IS COAL MADE?

Millions of years ago, the remains of plants and animals settled at the bottom of rivers and swamps. Water and dirt piled slowly on top. Heat and pressure forced out the oxygen as the material hardened and became coal. Today, miners extract the coal in layers from the ground. But coal is not renewable, so before we run out, we need to find new sources of renewable energy.

# AUSTRALIAN COAL

Australia is one of the largest sellers of coal to other parts of the world. Australia has only roughly 6 per cent of the world's coal resources and 2 per cent of its natural gas resources, but we have lots of mines.

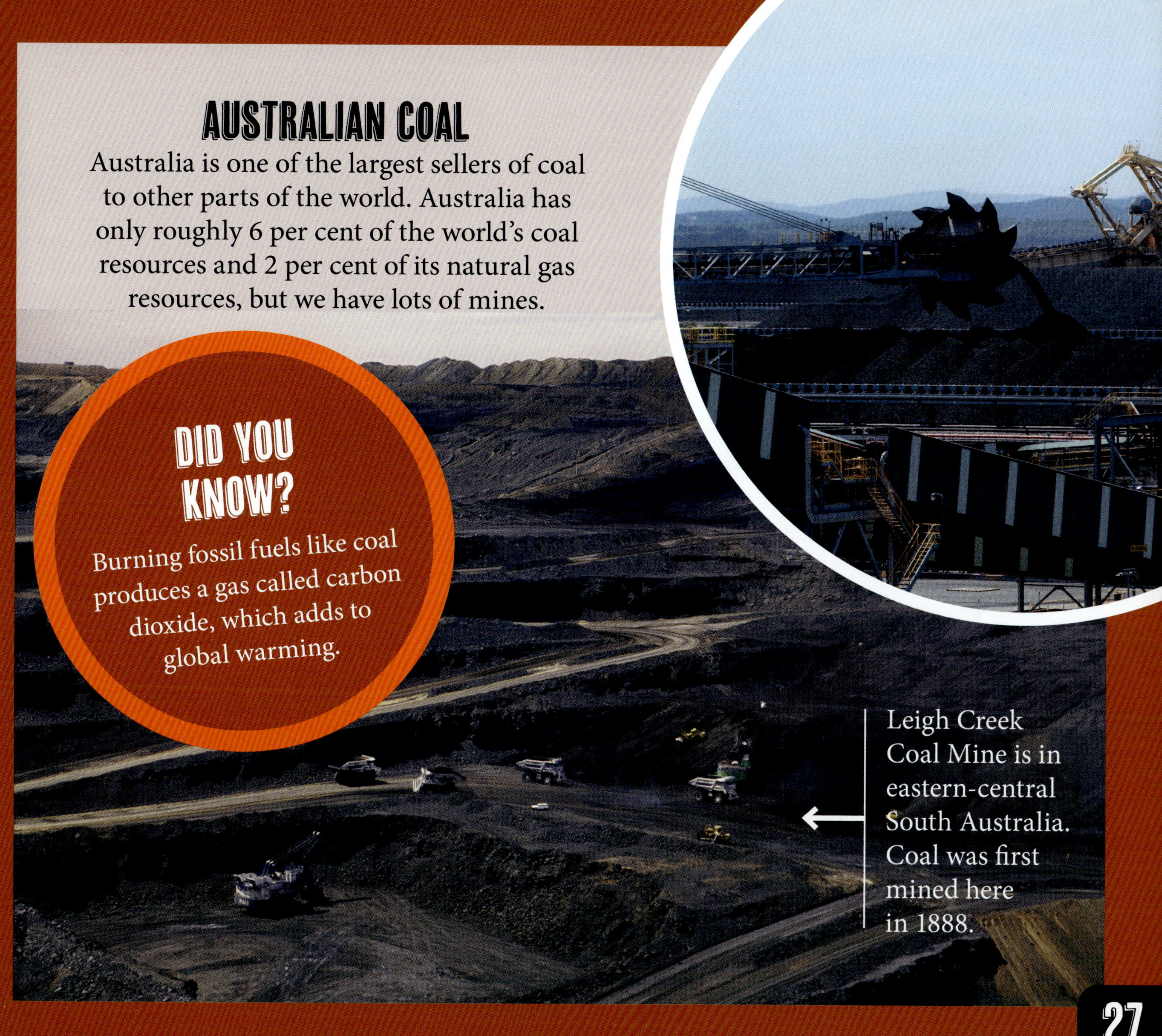

## DID YOU KNOW?

Burning fossil fuels like coal produces a gas called carbon dioxide, which adds to global warming.

Leigh Creek Coal Mine is in eastern-central South Australia. Coal was first mined here in 1888.

# LIVING FOSSILS

Stromatolites and thrombolites (which are similar, although thrombolites have a lumpy structure rather than a layered one) are rock-like living things that formed over 3.5 billion years ago. They harness energy from sunlight and produce oxygen in the process. They are thought to be responsible for increasing oxygen levels on Earth. These are some of the oldest known fossils.

## DID YOU KNOW?

Stromatolites were once the main form of life on Earth, and we have them to thank for the oxygen that allowed us and all other complex life to evolve!

LAKE CLIFTON, WA

## STROMATOLITE FOSSILS

Until 1961, stromatolites were only known to exist as fossils. Huge fossil sites tell us that stromatolites were once abundant in the Pilbara and other parts of Western Australia.

## LIVING STROMATOLITES

You can find living stromatolites and thrombolites in shallow lakes in Western Australia's Nambung National Park, in Hamelin Pool in Shark Bay World Heritage Area (north of Perth), and at Lake Clifton (south of Perth). The stromatolites in Shark Bay and Lake Clifton are roughly 3500 years old.

# GLOSSARY

**BACTERIA**
Tiny living things. They are the simplest living organisms.

**EXTINCT**
A species whose last living member has died.

**LAVA**
The molten or fluid rock (magma), which comes out of a volcanic vent.

**MAGMA**
A very hot, molten substance beneath the crust of the Earth.

**MINERALAS**
Solid, naturally occurring substances.

**OXYGEN**
A colourless chemical element that we breathe in and which helps many living organisms, including humans, to function.

**SEDIMENT**
Usually a mineral that has been moved by water, air or ice and settled somewhere in a layer.

**STAMPEDE**
A sudden rush of a group of animals, usually when they are spooked or frightened.

# PICTURE CREDITS

Images listed clockwise from top left unless specified otherwise: AG = Australian Geographic; SS = Shutterstock.com; US = Unsplash.com; CP = CanvaPro

**Front cover:** Bjoern Wylezich/SS; Warpaint/SS; Photo-world/SS; BThaiMan/SS; Catmando/SS; John Pickrell/AG; Xing Lida/AG; Vitaly Korovin/SS; YourElechka/SS; Byjeng/SS; Frances Mocnik/AG. **1:** Mark_Kostich/SS; Xing Lida/AG; TinaImages/SS; Sebastian Janicki/SS; Bjoern Wylezich/SS; Warpaint/SS; Ivan Smuk/SS; BThaiMan/SS. **2:** Vitaly Korovin/SS; Don Fuchs/SS; Marc Witte/SS; Quentin Chester/AG; Quentin Chester/AG. **3:** www.sandatlas.org/SS; Anna Dufour/SS; Kasin/SS; Only Fabrizio/SS; Pung/SS. **4:** Alan Pryke/SS. **5:** Taras Vyshnya/SS; Drew Hopper/AG; macrowildlife/SS; John Pickrell/AG. **6:** Chrissie Goldrick/AG. **7:** Drew Hopper/AG; Ralf Lehmann/SS. **8:** pixelrain/SS. **9:** Ra'ike/Wikimedia Commons; vvoe/SS; vvoe/SS; vvoe/SS; MarcelClemens/SS; vvoe/SS; vvoe/SS; vvoe/SS; vvoe/SS. **10:** Sebastian Janicki/SS; Yut chanthaburi/SS; Bjoern Wylezich/SS; Tycson1/SS; Bildagentur Zoonar GmbH/SS; Imfoto/SS; Imfoto/SS; photo-world/SS. **11:** TinaImages/SS; Potapov Alexander/SS; tanapat prompa/SS; Rory McGuinness/AG; Rory McGuinness/AG; Sebastian Janicki/SS. **12:** Nick Rains/AG; Jackson Stock Photography/SS; BThaiMan/SS; Jiri Vaclavek/SS; Gilles Paire/SS. **13:** John Pickrell/AG; Nick Rains/AG; John Pickrell/AG. **14:** Justin Walker/AG. **15:** Nick Rains/AG; Benny Marty/SS. **16:** Vitaly Korovin/SS; AG; Alex Coan/SS; Frances Mocnik/AG; Don Fuchs/AG; Frances Mocnik/AG. **17:** Linda Bucklin/SS; Peter Boer/Flickr; Bill Bachman/AG. **18:** paleontologist natural/SS; Warren Field/AG; Warren Field/AG. **19:** Warren Field/AG; Frances Mocnik/AG. **20:** Ivan Smuk/SS; Jake Kohlberg/SS; Jake Kohlberg/SS; Jake Kohlberg/SS; Mark_Kostich/SS; Jake Kohlberg/SS; arousa/SS; Jake Kohlberg/SS; Warpaint/SS. **21:** D. Wacey; Warpaint/SS; Catmando/SS; Akkharat Jarusilawong/SS. **22:** Frances Mocnik/AG; Bob Smith; David Lade/SS. **23:** Xing Lida/AG; lin padgham/Flickr; Xing Lida/AG; Matt Martyniuk/Wikimedia Commons. **24:** Anne Musser/Australian Museum. **25:** Vincent Long/AG; John Pickrell/AG; Carl Bento/Australian Museum. **26:** Vladyslav Trenikhin/SS. **27:** Mike Langford/AG; Beau Pilgrim/AG. **28:** Adwo/SS. **29:** Mike Langford/AG; David Bristow/AG; EA Given/SS. **30:** Sebastian Janicki/SS; Tycson1/SS; Imfoto/SS. **31:** Boyloso/SS. **32:** Warpaint/SS; Jake Kohlberg/SS.

**Back cover:** Drew Hopper/AG.

Australian Geographic

# DISCOVER


Australian Geographic *Discover: Rocks and Fossils* is published by Australian Geographic.




First published in 2019, reprinted in 2021, 2023
© Australian Geographic Holdings Pty Ltd
52–54 Turner St, Redfern, NSW

editorial@ausgeo.com.au
australiangeographic.com.au

**Text:** Natsumi Penberthy
**Managing editor:** Katrina O'Brien
**Managing commercial editor:** Lauren Smith
**Assistant commercial editor:** Rebecca Cotton
**Designer:** Mike Rossi
**Creative director:** Mike Ellott
**Print Production:** Andy Franks

AUSTRALIAN GEOGRAPHIC
Managing Director: David Haslingden
Licensing and Publishing Manager: Tom Bates
Commercial Assistant: Felicity McManus

Printed in China by C & C Offset Printing Co. Ltd.
The paper in this book is FSC® certified. FSC® promotes environmentally responsible, socially beneficial and economically viable management of the world's forests.

# BOOKS IN THIS SERIES

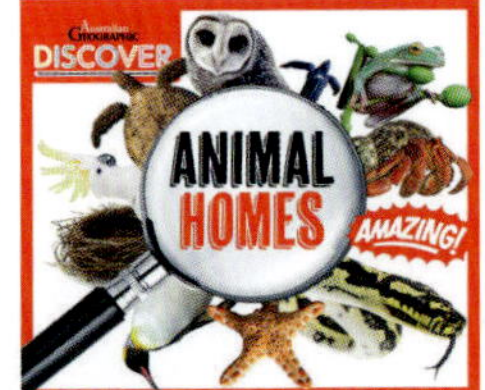


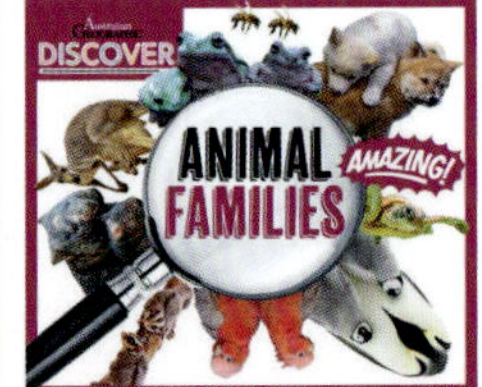

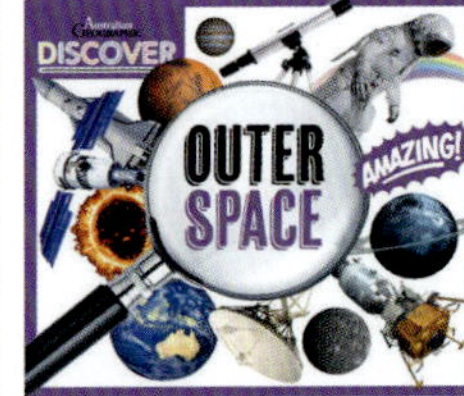

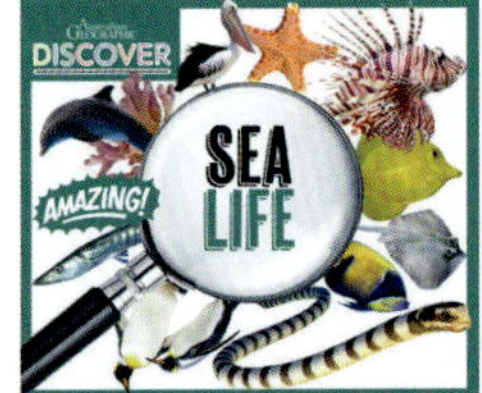


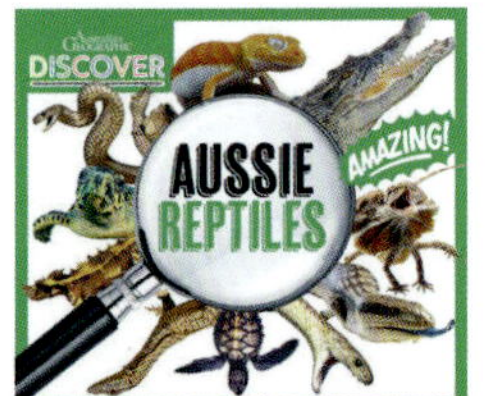


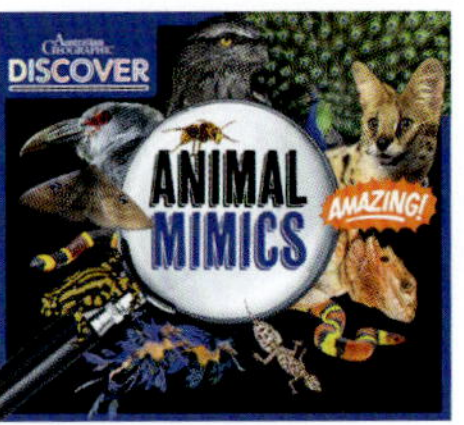


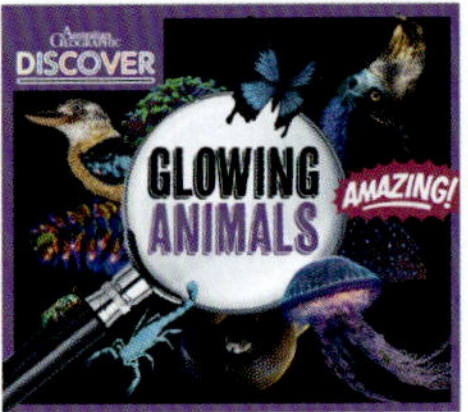

Australian Geographic contributes 100% of its profits to the Australian Geographic Society, including its conservation and sustainability programs.